# HUSKIES OF THE HEROIC ERA
# OF ANTARCTIC EXPLORATION

## David Jensen

For my daughter Gretchen and her children
Alexander and Tianna who all love animals

## THE ERSKINE PRESS

## 2018

Huskies of the Heroic Era
of Antarctic Exploration

First published in 2018 by
The Mawson's Huts Foundation.
Sydney, Australia

This edition published by
The Erskine Press, The White House, Eccles, Norwich NR16 2PB
WWW.ERSKINE-PRESS.COM

ISBN 978-1-85297-126-7

Typeset and design by Esme Power
Printed and bound in Great Britain by 4edge Ltd, Hockley.

# CONTENTS

The royalties from this book go directly to the Mawson's Huts Foundation to contribute to the conservation of Mawson's Huts at Cape Denison, the home of some of the huskies from the Australasian Antarctic Expedition of 1911-14.

The Foundation is a not-for-profit charity established in 1997 to conserve the historic buildings. At the time of publication the Foundation had funded and organised 13 major expeditions.

This work has saved Mawson's Huts from certain destruction and from being blown into the Southern Ocean. The Foundation works in close partnership with the Australian Antarctic Division.

Front cover photograph was taken by Mawson's official photographer Frank Hurley. It shows Xavier Mertz, one of the expeditions' dog-carers, training huskies at Cape Denison in 1912.

# 1

## INTRODUCTION

Huskies helped explorers and scientists in the Antarctic for over a century. They first stepped ashore in Antarctica on February 17, 1899 and left their paw marks on the ice for the very last time 95 years later in 1994.

During those years they made an incalculable contribution to science and exploration before the inclusion of an environmental clause in the Antarctic Treaty. This required the removal of all non-native species because of the threat of distemper, a dog disease, being spread to native seals.

Australia was one of the leaders in the use of "husky power" with over a thousand used at Australian bases for nearly 50 years. Scientists and explorers from several other countries also used huskies until the decision was made to withdraw them from the Antarctic.

The last Australian husky departed the Australian Antarctic Division's base at Mawson in September 1993 and those used by other nations a few months later.

These beautiful descendants from the wolves of Northern Asia played a major role in the exploration of Antarctica by being able to survive and work in extreme temperatures.

Just as importantly, they aided scientists in their research and science programmes with their ability to navigate crevasses and sea ice while hauling sledges laden with supplies and scientific equipment.

Dog-handlers believed they sensed hidden danger, particularly when moving over sea ice, halting in their tracks to alert their handlers if there was possible danger ahead from thin ice or tide cracks covered in snow.

Eskimos or Inuits (the North American word for "eater of raw meat"), are a group of culturally similar indigenous people who inhabit the Arctic regions of Alaska, Greenland and Canada. They first began breeding huskies as a hardy working animal 3000 years ago.

Naturally good hunters, because of having to survive mainly on seals, the dogs do not attack humans and quickly form a strong bond with their handlers.

Distant relatives of the wolves that roamed North Asia they got their name husky from English sailors in the 1850s who pronounced the name Eskimos as "Huskimos." The dogs they kept were given the name husky.

Husky today is the general name of the sled dog in northern regions. They are bred all over the world with the Alaskan Malamute the largest, most powerful and strikingly beautiful.

Alaskan Malamutes with Alaskan Inuits

Most of the huskies taken to the Antarctic came from Greenland and Labrador which produced the largest of the species bred by Inuits and was the source of those used during the "heroic era" of Antarctic exploration which began at the Sixth International Geographical Congress in London in 1895. The conference adopted the resolution "...that this congress record its opinion that the exploration of the Antarctic regions is the greatest piece of geographical exploration still to be undertaken". The "heroic era" ended in 1916.

Alaskan Malamutes were first bred by the Mahlemiut people who lived in the upper western region of Alaska. This was part of Russia before it was sold to the United States for $7.2 million in 1867.

While out hunting for food and survival, Malamutes were used to look after the Mahlemiut children, which demonstated their gentle nature with humans.

Not all the famous names of Antarctic exploration were convinced of their value. Those who did included the Norwegian Roald Amundsen and Australia's own Douglas Mawson. Amundsen credited his successful drive to be the first person to reach the South Pole to the use of huskies. Huskies also served Mawson with distinction and saved his life during his epic trek for survival.

The British explorers Robert Falcon Scott and Ernest Shackleton did not share their enthusiasm. Had they done so history may well have been different. Both men took huskies with them on their expeditions but preferred ponies and they experimented with motor vehicles, both of which were considered failures.

Shackleton walked to within 180 km of the South Pole during his 1907-09 Nimrod Expedition, using Manchurian ponies in the early stages of the trek. However he was forced to eliminate them during the journey and a shortage of rations prevented him from reaching his goal.

He turned for home knowing he and his men would not have sufficient food to survive if they continued. As he later told his wife, "Better a live donkey than a dead lion." Historians now believe Scott could also have returned safely after reaching the South Pole had he used properly trained dogs but the Englishman preferred to do it the hard way, man-hauling sleds, regardless of the circumstances.

The first dogs to set foot in the Antarctic were part of the British Antarctic Expedition (the Southern Cross Expedition) of 1898-1900. They were put ashore at Cape Adare and immediately showed their value when seven of the crew of the *Antarctica* were trapped on the ice by a blizzard for four days which interrupted the landing. The men survived by sleeping with the dogs for warmth.

The use of huskies in the Antarctic, particularly during the "heroic era", was not always pleasant. Both men and dogs were fighting for survival in the world's harshest environment and both suffered. Sacrifices had to be made for human survival, despite dogs and men fashioning close relationships.

This book focuses on huskies used during the "heroic era" of Antarctic exploration. During that period the pioneers and giants of early exploration of the white continent were at the time household names.

The adventures and exploits of Amundsen, Scott, Shackleton, Mawson and Borchgrevinck gripped public imagination at the time. The Scott tragedy and Mawson's survival along with Shackleton's extraordinary story still rank among the greatest ever stories of Antarctic exploration.

Huskies played a major role in the early exploration of Antarctica and it's important that their part in this period of Antarctic history is understood and remembered.

It is fitting that their work for Mawson has been immortalised by the

Australian Antarctic Division's Place Names Committee which in 2017 announced over 20 prominent landmarks would be named after some of huskies.

Their colourful names will forever play an important role for international scientists, researchers and rescuers as they navigate their way across the white continent, Each name is tied to a distinct geographic feature.

This will help ensure these remarkable animals will be remembered.

Stareek and Osman - two huskies with Scott

## 2

## FIRST STEPS IN ANTARCTICA

The use of huskies in the Antarctic was pioneered by the British Antarctic Expedition of 1898-1900. Also known as the Southern Cross Expedition, it was led by the Anglo-Norwegian, Carsten Borchgrevink.

A school teacher with a Norwegian father and English mother, Borchgrevink moved to Australia in 1888. In 1895, after teaching in New South Wales, he joined a commercial expedition on board the whaler *Antarctica* which sailed into waters named after the ship and landed at Cape Adare. He was just 24.

Borchgrevink giving three cheers for Sir James Clark Ross
on Possession Island in 1895.
From *The Strand Magazine, 1897*

Skippered by Henryk Bull, a party consisting of himself and Borchgrevink laid claim to be the first to land on the continent, though an American sealer John Davies claimed he stepped ashore on the Antarctic Peninsula in 1821.

On return to Australia Borchgrevink strived to fund his own expedition and eventually secured money from Australian and British investors. He eventually found a sponsor on the understanding it was labelled a British expedition.

The dogs for this expedition were found in Siberia by a member of the Royal Geographical Society, Dr J Russell Jeaffreson, an explorer who volunteered to find the animals required.

He travelled 1300 kms to Siberia where Samoyed dogs were bred by the Samoyedic people for herding reindeer.

Having obtained a pack of them for the expedition, he returned to England where they multiplied in numbers and attracted great interest.

The Expedition landed at Cape Adare on February 17, 1899. After a three week stay in Hobart following the voyage from London, Borchgrevink's 75 huskies became the first dogs to step onto the continent.

Looking after them were two Norwegian Laplanders, Ole Must and Persen Savio, who wrote themselves into history books by becoming the first to drive dog teams in the Antarctic.

Another member of the team was also the first Australian to winter in the Antarctic.

He was Louis Bernacchi, a Tasmanian, who was placed in charge of meteorological observations and photography. His diary notes were published in "That First Antarctic Winter: the story of the Southern Cross expedition of 1898-1900," written and edited by his granddaughter, Janet Crawford.

On arrival in the Antarctic the men survived their first night ashore in a large tent by getting all the dogs to lie on top of them for warmth. One of the huskies was blown out to sea on an ice floe but turned up 10 weeks later in good spirits, showing huskies could survive in the world's coldest continent. Kennels were made for them out of packing cases.

It was the first expedition of the heroic era of Antarctic exploration and also the first to spend a winter on the continent, making Borchgrevink a true pioneer of Antarctic exploration.

The wooden huts they erected in the middle of a penguin colony were named "Camp Ridley". They remain standing today and are looked after by the NZ Antarctic Heritage Trust.

Bronze of Louis Bernacchi with husky, Hobart waterfront.
Photo courtesy of David London

# 3

## HUSKIES CONQUER THE SOUTH POLE

Roald Amundsen

Norwegian Roald Engelbregt Gravning Amundsen did not hesitate to use huskies after deciding to be the first person to reach the South Pole.

He ordered 100 North Greenland sledge dogs which were considered the strongest and best. He could not understand the apparent reluctance or aversion by Scott and Shackleton to using dogs.

*Can it be that the dog has not understood its master? Or is it the master who has not understood the dog?* Amundsen later wrote.

Amundsen with huskies

The Norwegian often referred to his dogs as his "children" and told his men

*the dogs are the most important thing for us. The whole outcome of the expedition depends on them.*

Unlike Scott, the Norwegian took them in the knowledge that some would be sacrificed for food if required. Only 11 returned from the expedition.

Before leaving for the Antarctic, Amundsen was thorough in his preparations spending time in the Arctic studying how Eskimos lived, the clothing they wore and how they used their dogs.

In comparison, Scott spent time in the Scottish Highlands with his ponies.

Amundsen also kept details of his attempt to reach the South Pole a closely guarded secret.

Only his brother knew the true story. Scott had thought Amundsen was planning an expedition to the North Pole and had sent

Oscar Wisting and his dogs at the South Pole
December 14, 1911

instruments to him in Norway in the hope that comparative readings could be made at opposite ends of the earth. That move created awkwardness. When Scott was in Norway testing motor sledges for his expedition he telephoned Amundsen at his home to discuss possible co-operation but his call was not taken.

Scott was aware that Amundsen was using huskies for his expedition, supposedly to the North Pole.

When Amundsen departed Norway not even his crew knew his real intentions. They had been told originally they were heading for the Arctic but doing so via Madeira. It was only after leaving the Portuguese archipelago in the North Atlantic Ocean that he revealed his plans.

They sailed south and Amundsen established his Antarctic base in January 1911 on the shores of the Bay of Whales in the Ross Sea. He called it Framheim (Home of the Fram) after his ship the *Fram*.

His dogs spent the first few weeks of their time in Antarctica hauling timber and supplies from the ship to their new home, erected four km from the coast line.

Planning was meticulous, with teams of dogs used to establish a series of food depots along the planned route to the South Pole. These supplies included 200 tonnes of seal meat for the dogs.

So intense was the cold in March and April that several of the huskies died along the way but Amundsen was still full of praise for them and again questioned the British aversion to using them.

At that time Scott was establishing his own base nearby on Ross Island at McMurdo Sound and while the two leaders did not meet, several members of their expeditions did.

Scott's ship the *Terra Nova* had dropped their leader off at McMurdo

Sound after arriving from New Zealand and a small party then sailed on to explore the area. They unexpectedly encountered *Fram* and were invited to dine on board.

It was then that Amundsen was relieved to be told Scott did not have wireless which could have announced to the world the Norwegian's plans to be the first to reach the Pole.

On October 19, 1911 the Norwegians left Framheim and began the trek to the South Pole taking 52 dogs, four sledges and five men. By mid-November when they had reached the top of a glacier surrounded by the Trans-Antarctic Mountains seven dogs had died.

Preparing for the final trek Amundsen selected 18 dogs which would take him to the South Pole and hopefully back to base.

He instructed each of the men in charge of sledges to kill the remainder. Amundsen wrote later:

*We called the place the Butchers' Shop, There was depression and sadness in the air; we had grown so fond of our dogs.*

Amundsen's South Pole Party, 1911

But their regrets did not prevent the team from enjoying the plentiful food the dead huskies provided and some of it was taken as rations for the remainder of the journey.

By December 4 they had reached the point where Shackleton had been forced to end his attempt three years earlier – just 180 km from the South Pole.

Ten days later with five men and 16 dogs still alive, Amundsen planted the Norwegian flag at what he called Polheim "home of the Pole".

He left a small tent and inside a letter addressed to the King of Norway explaining the team's accomplishment. Also left was a note for Scott asking the Englishman to deliver his letter to the King noting his team's accomplishment in case they failed to return. The note and his letter to the King were later found on Scott's body.

The journey home began while Scott's party was five weeks away from reaching the Pole.

Fram, Hobart harbour, March 1912
(Photogarph by JW Beattie)

Amundsen and his four men walked into their base with 11 surviving dogs on January 25, 1912. They had covered 3440 km, conquered the South Pole and taken less than 100 days.

After loading their supplies, equipment and remaining huskies they sailed for Hobart where they arrived on March 7. After rowing ashore Amundsen then walked up Macquarie Street, stared at by Hobartians, and booked himself into Hadley's Hotel in Murray Street, then the city's premier establishment.

Hadley's Hotel, Hobart – c. 1912

It was the first contact Amundsen had had with the outside world for two years and his appearance was not that of their usual guests seeking accommodation.

Somewhat reluctantly staff found a smallish room at the rear of the hotel. However this quickly changed the following day after he called a press conference on the steps of the Hobart Post Office.

His fame quickly spread through Hobart following a telegram to his brother asking him to inform the King of Norway that he had been successful in reaching the South Pole.

The news raced around the world and following his announcement Hadley's staff quickly upgraded their guest who had since also changed into more formal attire.

The "Amundsen Suite" is now quite sought-after by guests. The friendly Norwegian stayed for a further two weeks enjoying Tasmanian hospitality. His men were also given VIP treatment.

At that time Scott was just days away from his life being taken by cold and starvation. The Englishman and his four companions had arrived at the South Pole on January 18 to find Amundsen's tent and the Norwegian flag fluttering in the polar wind.

After photographs were taken Scott's party began their return, hauling their sledges which contained nearly 100 kg of rock specimens that they had collected along the way. He made his last diary entry on March 29 about a week after Amundsen had sailed from Hobart for Norway and a King's welcome.

Also moored in Hobart at that time and not far from the *Fram* was Mawson's ship *Aurora*. It had just returned from deploying Mawson's expeditions at three bases in the Antarctic and was preparing for scientific work in the Southern Ocean before returning to the Antarctic to collect Mawson and his team.

The 11 dogs which had taken Amundsen to the South Pole were generously donated to Mawson's relief expedition. They subsequently returned to Antarctica where they spent another year before arriving back in Australia in February 1914.

# 4

## DOGS, TRACTORS AND PONIES

British explorer Captain Robert Scott took huskies on his first Antarctic expedition in 1901-04 and his ill-fated return there in 1910-13 which cost him his life along with five companions.

On both occasions Scott and his men failed to understand how to train and look after their dogs. They were treated more as pets than as a means of transport and possible survival.

During the National Antarctic (Discovery) Expedition of 1901-04, his first, the party took 25 dogs but proper training had not been sought and the dogs were given the wrong food, including Norwegian dried fish, which was sometimes rotten and made them ill.

Scott's motor sledge

When the dogs lost condition and became weak and lame they limped along behind men who finished up hauling the sledges themselves. When they were fit they travelled so fast that the men, inexperienced with using skis, could not keep up. Heavier loads were used on the sledges to slow them down but with little improvement to the overall operation.

Almost the entire team of dogs on Scott's first expedition perished or were killed, with some eaten, the men having to resort to hauling the sledges themselves.

During the 1910-13 expedition Scott again took dogs but again failed to use them properly.

For transport on this expedition Scott preferred ponies and motorised vehicles. One of the tractors broke down 80 kms from base camp. For the ponies he had special snow shoes made but in soft snow they sunk up to their bellies.

Oates and some of the Manchurian ponies in his charge,
on their way to McMurdo Sound, 1911

Unfortunately the man Scott put in charge of the ponies was reluctant to fit them with specially made boots and his leader did not insist on them being used.

Antarctic notes – c. 1911 at McMurdo Sound
(Photograph by Herbert Ponting)

The explorer had planned to use ponies for the first quarter of his journey to the Pole; his approach being supported by the fact Shackleton had used ponies for part of his attempt in 1909 when he got to within 180 kms of his goal.

Being aware that Amundsen planned to use huskies, Scott wrote to his wife Kathleen from base camp:

*I don't know what to think of Amundsen's chances. If he gets to the Pole it must be before we do, as he is bound to travel fast with dogs, and pretty certain to start early.*

Before leaving Britain, Scott sought, and then ignored, advice from several experienced Polar explorers including another Norwegian, Fridtjof Nansen, who had used dogs to explore the Arctic. *...take dogs, dogs and more dogs* he advised Scott.

Scott believed dogs should only be taken if they could be returned home:

*I am inclined to state my belief that in the Polar Regions properly organised parties of men will perform extended journeys as well as teams of dogs. On the other hand, if the lives of the dogs were to be sacrificed, then 'the dog-team' is invested with a capacity for work which is beyond the emulation of men. To appreciate this is a matter of simple arithmetic.*

Scott simply refused to use dogs if they were to be slaughtered for food later.

*One cannot calmly contemplate the murder of animals which possess such intelligence and individuality, which have frequently such endearing qualities, and which very possibly one has learnt to regard as friends and companions.*

The Englishman also did not appear to consider that the food for the ponies had to be taken all the way from England while dogs could be fed on seals and penguins hunted in the Antarctic. While the British frowned on the Norwegians eating their dogs, Scott and his men ate their ponies.

Scott's ponies hauling supplies

Historians later discovered that Scott has left instructions that supporting dog teams were to meet and assist his party on the return trek. These instructions, which he dated October 11, 1911, were never implemented and were only discovered in the closing stages of the 20th century when commentators were questioning his competence and character.

Scott and his four companions reached their objective on January 17, 1912 only to find a small tent and the Norwegian flag flying from a bamboo stake. Roald Amundsen had beaten him to the Pole 32 days earlier and at that time was just one week away from reaching the safety of his ship using huskies to pull their sledge of supplies.

Scott and his men, struggling to pull their sledge, were just 15 kms from the nearest food depot and 241 kms from base camp when they perished.

The Englishman had a great fondness for his dogs and he wrote about them in his diary before departing Lyttleton, NZ where they provisioned the ship before sailing for the Ross Sea.

*Upon the coal sacks, upon and between the motor sledges and upon the ice-house are grouped the dogs, thirty-three in all. They must perforce be chained up and they are given what shelter is afforded on deck, but their position is not enviable. The seas continually break on the weather bulwarks and scatter clouds of heavy spray over the backs of all who must venture into the waist of the ship. The dogs sit with their tails to this invading water, their coats wet and dripping. It is a pathetic attitude, deeply significant of cold and misery; occasionally some poor beast emits a long pathetic whine. The group forms a picture of wretched dejection; such a life is truly hard for these poor creatures.*

There were a few survivors of Scott's huskies from that fatal expedition, but he would have been pleased that Osman, a favourite of his, finished his days in New Zealand's Wellington Zoo.

Scott's huskies training in Lyttleton Harbour.
(Photograph courtesy Canterbury Museum, Christchurch)

On board *Terra Nova*, 33 Siberian sledge dogs. Scott noted that the animals ...*had the most wretched time, even in calm weather.* One was strangled by his chain in a gale, another, Osman, was washed overboard by a wave, but was swept right back aboard.

Dimitri Gerov, the Russian dog-handler engaged by the expedition was presented with Osman. But when he returned home to Russia three years later Osman was given to a friend of Scott's, Joseph Kennedy who was NZ's representative on the expedition.

Osman had led a charmed and interesting life. Recruited from Siberia where he was used by the Russian postal service delivering mail across the frozen Sea of Okhotsk, he and the other 29 members of Scott's husky team spent 10 weeks on Quail Island in Lyttleton Harbour from where the 1910 expedition departed.

The dogs hauled makeshift sledges with very small wheels to traverse the snowless environment on the island.

In 1916 Osman was presented to the Wellington Zoo. He was referred to in the newspapers as "Osman the Great" and joined Shackleton's dog, Oscar, also a veteran of the Antarctic who had also been given a home there.

It is believed Osman died in early 1918 and was probably around 10 years old, the average life expectancy for a husky though not for one who had led such an adventurous life.

# 5

# SHACKLETON'S TWO ANTARCTIC EXPEDITIONS

Shackleton's Athol-Johnson

Ernest Shackleton took dogs and Manchurian ponies on his first Antarctic expedition in 1907-09. He also became the first person to take a motorised vehicle to the continent.

The vehicle constantly overheated and sunk into the snow and was of little practical use. Neither were the ponies.

They did not travel well on the final leg of their journey from New Zealand to the Ross Sea area from which Scott also used as his temporary base. One pony had to be shot on arrival at the ice.

*Endurance* being crushed
(Photograph Frank Hurley)

Shackleton used the ponies in his attempt to reach the South Pole, but not with great success. One of his men was kicked in the knee which exposed the bone, so deep was the gash. The animals were later killed for food, with the men also eating the maize taken to feed the ponies.

Learning from these experiences was a young Australian geologist, Douglas Mawson, and another Australian, Sir Tannant William Edgeworth David. Both had been selected to join Shackleton and members of the party which successfully reached the south magnetic pole and made the first ascent of Mt Eerebus.

Edgeworth David commented later he felt the journey to the magnetic pole could have taken half the time had dogs been used.

Mawson also took note of the value of the dogs when preparing for his expedition two years later.

During Shackleton's second expedition, the Imperial Trans Antarctic Expedition 1914-17 (the *Endurance* Expedition), he took additional dogs which helped save the lives of his men when *Endurance* became stuck in the ice and was crushed.

It was to be the last expedition of the heroic era with Shackleton looking to be the first person to make a land crossing of the continent.

Amundsen had beaten everyone to the South Pole but Shackleton planned to be the first to cross the continent, marching from Vahsel Bay in the Weddell Sea to the Ross Sea via the South Pole, using a team of six men and 70 huskies.

Shackleton required two ships for this expedition. *Endurance* took him and his main party to the Weddell Sea while the second ship, Mawson's old vessel, *Aurora.* would deposit a party of eight men and 18 dogs in the Ross Sea. Their task was to lay depots for Shackleton's party coming via the South Pole.

It was a disastrous expedition from the start for both ships. *Endurance* became stuck in the ice before reaching Vahsel Bay and was eventually crushed with huskies and crew watching.

In the Ross Sea there was also disaster. The party arrived in January 1915, later in the summer than scheduled. The men and 18 dogs were not acclimatised to the icy conditions.

Their first journey to lay depots saw the loss of 10 of the huskies and only one depot laid.

In May that year *Aurora* had been blown from its moorings leaving 10 men and seven dogs stranded on the ice with many of their supplies still on board.

The ship itself was caught fast in the ice and drifted 2600 kms over nine months before breaking free. It eventually limped its way back to NZ but the men left behind had to remain there until they were rescued in January 1917.

The laying of the depots was completed late in 1915 but to no avail as Shacklelton's party had to abandon their ship. Just before it disappeared photographer Frank Hurley, fresh from Mawson's expedition, dived into the freezing waters to save valuable glass plate negatives.

Shackleton, however, only allowed the Australian to take a small number of slides and equipment with him because of the need to minimise weight.

On *Endurance* were 28 men and 69 dogs, The dogs were mostly of mixed breed and each weighed about 45 kgs. All crew members were assigned a husky to look after and strong bonds were formed during their time drifting on the ice.

Their names included Bummer, Chips, Hercules, Judge, Roy, Samson, Satan, Shakespeare, Slippery Neck, Steamer, Stumps, Surly, Swanker, Upton and Wallaby.

They lived in "dogloos" built from ice and wood. Sadly all had to be killed by their handlers when the sea-ice broke up and Shackleton made the decision to row to Elephant Island in three boats.

Once there Shackleton and five others rowed 1287 kms in open sea to South Georgia to find help. Which he did. After rescuing his men from Elephant Island, Shackleton then focused on retrieving the Ross sea party early in 1917. Sadly, only seven men and five dogs survived.

On his return to New Zealand, Shackleton tried to sell the dogs at an auction in Wellington and one, Oscar, joined Scott's favourite husky, Osman, at the local zoo.

Tom Crean with puppies on Endurance, 1911
(Photograph Frank Hurley)

# 6

# DOUGLAS MAWSON

Douglas Mawson

Douglas Mawson always favoured the use of huskies and debated this with Shackleton during his time as a member of the British explorer's 1907-09 expedition.

Shackleton preferred using Manchurian ponies which, although being able to haul over a ton of supplies, were harder to replace than dogs if lost down a crevasse.

The Australian was quite passionate about the use of huskies and his wife Paquita wrote in her book *Mawson of the Antarctic* that her husband had a strong feeling in favour of dogs:

> *He liked animals and I have an idea he liked the way the dogs used to run forward, wagged their tails and stood waiting to be harnessed into their places in the sledge team. You do not get a pony to do that.*

*In later years at Douglas's lectures ladies were sometimes turning to one another to register disapproval when Douglas mentioned dogs being used to pull sledges. I told him this once and afterwards, when he thought about it, he would pause for a moment, look at his audience and remark how pleased the dogs were to be working again and how they would bark, eager to be off.*

In his official diary *Home of the Blizzard* Mawson wrote:

> *There can be no question as to the value of dogs in the Polar regions, except when travelling continuously over very rugged country, over heavily crevassed areas or during unusually bad weather.*

*It is in such circumstances that the superiority of man-hauling has been proved. Further in an enterprise where human life is always at stake it is only fair to put forward the consideration that the dogs represent a reserve of food in case of extreme emergency.*

Mawson would have perished had he not exercised an "extreme emergency" situation in using huskies as a food reserve.

One of Mawson's first moves when planning for the 1911-14 Australasian Antarctic Expedition (AAE) had been to select a team of 50 Greenland huskies.

He appointed a young British army officer Lt Belgrave Ninnis and a former Swiss ski champion, lawyer and mountaineer, Xavier Mertz to look after them.

On-board his expedition ship, *Aurora,* that Ninnis and Mertz sailed on from London on July 27, 1911 were 47 huskies which Mawson referred to as "Eskimaux dogs". They were bound for Hobart, some of them already in pup.

Sadly they lost one before leaving British waters. Just before reaching Cardiff to take on coal, Ninnis wrote in his diary:

*We had the misfortune today to lose a bitch, our first death among the dogs. We were particularly unfortunate in losing this one as we were daily expecting a litter. However, we have our two pups in a sound and healthy state.*

*Yesterday Mertz was caught in the act of taking "Pavlova"* (they had already named the dogs) *away from the ship for a walk in Cardiff. The penalty for landing dogs which have not been in quarantine is I believe £500 sterling so the little jaunt, if accomplished, would have cost as much as the coal.*

A hut was built on the deck of *Aurora* for shelter but it was not totally effective. The heat and rough seas took their toll, particularly of pups born. Since leaving Greenland the dogs had been grouped together and pregnancy seemed contagious.

Over 30 pups were born in one week just after leaving Cardiff and Mertz and Ninnis had their hands full. Most did not live for long. Nineteen perished in just one day from heat and others were washed overboard.

McMurdo Strait - The Discovery in winter quarters, frozen in since March 1904. Photographed by Lt Shackleton and included in *The Illustrated London News,* March 30 1912.

Dogs at the Grottoes — Alexander Kennedy's diary
Mawson's AAE — 1911-14

A Siberian Husky with Bernacchi on the Southern-Cross/ Borchgrevink expedition

Watson and Sandow— Andrew Watson's diary Mawson's AAE 1911-14

Ninnis hauling up a slope

Erebus—a samoyede—Shackleton's Trans Antarctic Expedition

Drawn by a member of Shackleton's *Nimrod* expedition and used in
the 1913 edition of ANTARCTIC DAYS

Loading huskies at Deception island (courtesy the Russell family)
Dog-sledging—Hope Bay (courtesy Justin Marshall)

In 1944 nine men landed on a tiny island off the Graham Land Peninsula. Armed with a small assortment of rifles and pistols and an obsolete 12 pounder mounted on the bow of their decrepit supply vessel they were ostensibly there to prevent German U-boats and surface raiders from using Antarctic harbours, but their real purpose was to reassert British territorial rights in the face of increasingly confident incursions by neutral Argentina. This two year expedition was code-named Operation Tabarin.

These pictures are taken from TWO YEARS BELOW THE HORN—A Personal Memoir of Operation Tabarin by Andrew Taylor, edited by Stephen Haddelsey. (Erskine Press 2017)

THIS SCULPTURE HONOURS HUSKIES
THAT SERVED EXPLORERS AND
SCIENTISTS IN ANTARCTICA BETWEEN
1898 AND 1994. BASILISK (STANDING)
AND ALEXANDRA (ON SLEDGE)
WERE WITH THE AUSTRALASIAN ANTARCTIC
EXPEDITION (AAE)1911-14 BASED AT
MAWSON'S HUTS, CAPE DENISON, EAST ANTARCTICA.

POSTCARDS:
Shell: advertising card — issued about 1910;
Fry's Cocoa: advertising card — issued about 1910
Joyce, dogs and penguins — probably issued late 1909
Shackleton's ponies arriving in the Antarctic — probably issued late 1909

Top: Huskies would have been better.
Shackleton's *Nimrod* Expedition

Left: Mawson and Ninnis

Right: Collecting penguins
Mawson's AAE

*Endurance,* shortly after leaving South Georgia, showing the arrangement of kennels.

Frank Wild and friends on *Endurance*

A veterinarian boarded the ship in Cape Town and diagnosed distemper for at least two of the bitches who died soon after leaving South Africa. On arrival in Hobart in November they were put into the Nubeena Quarantine Station at Taroona, just outside the city.

By December just 36 dogs of the original 47, and two pups, remained.

A further eight died before the ship departed for Macquarie Island and by the time the AAE reached Cape Denison, where they established the main base, the number had dwindled to just 28. One had been left behind on Macquarie Island because it was too aggressive and others had died from various causes.

Of the rest, 19 were to remain at Cape Denison with 18 men. Nine were taken by the eight members who formed the Western Party, under the leadership of the legendary Frank Wild, which established its base on the Shackleton Ice shelf.

Huskies enjoying time on Macquarie Island
(Photograph Frank Hurley- courtesy Museum of South Australia)

AAE's Western party unloading huskies and supplies – Shackleton
Ice Shelf, February 1911
(Photograph courtesy Mitchell Library, NSW)

Yorkshire born Wild had met Mawson when the two men were members of Shackleton's 1907-09 British Imperial Antarctic Expedition. Wild had already been south with Scott in 1901-04 and at the age of 39 was highly regarded as a veteran Antarctic explorer.

The nine dogs in that party did not play a major role. They were Sandow (named after Eugene Sandow, the world's strongest man in that era), Tich (originally Little Tich, an English music hall comedian), Switzerland (named by Mertz), Amundsen (Roald Amundsen) Sweep (members of the AAE regularly ran betting sweeps), Tiger (had black and white stripes), Zip (nickname of one of Ninnis' closest friends), Crippen (an English murderer who was hung near the time of the expedition) and Nansen (Norweign Arctic explorer).

Two died soon after arriving, two disappeared and two more were shot to save rations for the remainder.

Of the remaining three, one had to be shot when the task of hauling sledges became too difficult for it. The remaining two, Zip and Amundsen, also known as "Chucklehead", finally returned safely to Australia.

Zip, who was the favourite of the Western Party, saw out his final days hauling tourists on a sledge at Mt Kosciusko. Amundsen died while waiting in quarantine at Taroona near Hobart.

Mawson's team of dogs at Cape Denison fought, bred and hauled supplies for the men who undertook several sledging journeys in November-December-January of 1912-13.

Sixteen huskies were involved on Mawson's epic trek which cost the lives of Ninnis and Mertz and from which Mawson barely survived. The party left the main base on November 10, 1912, along with Blizzard, the pup born at Cape Denison. None survived.

Frank Hurley cutting dog rations, Cape Denison, 1912
(Photograph courtesy Mitchell Library of NSW)

During the next few weeks the three men soon sorted out the best dogs and there were several near disasters, one of which demonstrated Mawson's incredible strength. The Australian almost single-handedly saved a whole team of six dogs from plunging down a crevasse.

Mawson was controlling two sledges joined with six dogs when the animals fell. With the dogs struggling in their harnesses the explorer dug in his heels to hold them and the two sledges while Ninnis raced for a rope to help haul them to safety.

Slowly Mawson reeled them back from almost certain death until he and Ninnis had them back on firm ground. But minutes later two dogs, who had been released from their harness as they could not work, were in danger.

Desperately clawing at the edge of a broken snow bridge Ginger Bitch, (also known as Alexandra), and one of her offspring, Blizzard, were seconds from death.

Ginger Bitch was in late pregnancy and Blizzard had an injury to a paw after being struck by a sledge.

Mertz managed to grab Blizzard and Mawson caught Ginger Bitch by her hackles. She celebrated by giving birth to a litter of 14 pups, all of whom died within minutes.

Blizzard, who had become the hut favourite had developed into a large dog and was the only pup to be born in the Antarctic and develop into a working dog.

While all this was happening Ginger (named because of its colour) wriggled free of its harness and disappeared back the way they had travelled. Days later he returned and was found in the morning lining up with the other dogs for a meal without a care in the world.

During the next week Blizzard had to be euthanised as he could no longer haul a sledge because of his injured leg. That left 12 dogs.

The three men were only two days from turning for home when disaster struck.

On December 12, 1912 Ninnis plunged to his death, along with of the dogs. The dogs were Basilisk (mythical king of serpents), Alexandra, Shackleton, John Bull, Castor and Franklin. They were also the strongest and had expected to be the main haulers on the 500 km trek home.

Ninnis' premonition he had in London prior to the expedition, that he would fall into a crevasse, became reality. Because of this he had secreted in his pocket a cyanide pill to take if he was injured in a crevasse and could not be rescued.

He would not have had time to take it. The crevasse Mawson and Mertz peered into for a sign of life was dark and deep. On a ledge 48 m below were two dogs, one dead and the other injured, whining soon to fall silent. Beyond that there was just darkness.

Gone with their good friend Ninnis were precious food, fuel, their tent and their six best dogs.

Mertz and Mawson were left with George (King George V), Mary (Queen Mary), Johnson, Haldane (UK Secretary of State when Ninnis was in the fusiliers), Ginger and Pavlova.

As the food ran out the emaciated dogs were shot and fed to their mates with some pieces going to the men. They had all died by the time Mertz passed away leaving Mawson to stagger on alone to eventual safety.

Mawson and Mertz had carried some of the dogs on their sledge when they became too weak to walk but eventually they were all killed to help the survival of their masters.

It was the loss of supplies down the crevasse, which also took all the food reserved for the dogs, that exacerbated the problems for Mawson and Mertz.

Mawson wrote in his diary:

*As we worked on a system which aimed at using up the bony parts of the carcase first, it happened that Ginger's skull figured as the dish for the last meal. As there were no instruments capable of dividing it, the skull was boiled whole and a line drawn around it marking it into right and left halves. These were drawn for in the old and well-established sledging practise of "shut-eye" after which, passing the skull from one to the other, we took turns in eating our respective shares. The brain was certainly the most appreciated and nutritious section. Mertz, I remember well, remarking specially upon it.*

Mawson explains:

*In explanation of the term "shut-eye" it should be stated that on sledging journeys it is usual to divide the all food in as nearly even portions as possible. Then one man turns away and another, pointing to the share, and addressing the former asks, "whose?" The reply is "yours" or "Mine". In this way an impartial and satisfactory division is made.*

Scientists believe the men were poisoned from eating the livers of the dogs as those of Greenland dogs contained toxic amounts of Vitamin A.

Mertz succumbed to the poisoning early on January 8, 1913 leaving Mawson in a terrible state and still over 100 km from base. He kept painfully and slowly marching on, growing weaker by the day.

When only nine kms from the hut he was forced, due to a blizzard, to seek refuge in a snow cave. A week later, on February 8, he staggered to within sight of the hut; he had been 32 days alone on the edge of survival.

Six colleagues who had stayed behind for another year to search for survivors or bodies rushed to meet him. *Good god, which one are you* were the first words uttered.

Mawson eventually left Cape Denison nearly a year later with his companions and 11 dogs bound for Adelaide to be re-united with Paquita.

Top: Some of the dogs on board *Aurora*
Left: Belgrave Ninnis in his Royal Fusiliers uniform
Right: Mertz demonstrating his climbing abilities on an ice cliff

# 7

# LAUGHTER, EXCITEMENT, MERRIMENT, DEVOTION AND DEATH

Belgrave Edward Sutton Ninnis and Xavier Mertz, the two men appointed by Mawson as the expedition's dog carers, were both novices for this important task. But they loved their huskies and Mawson could not have allocated the task to better men.

Polar exploration was in Ninnis' blood. His father was a member of a British Arctic Expedition in 1875-76 and a cousin was a member of Shackleton's Ross Sea party in the 1914-17 Imperial Trans-Antarctic Expedition. He had also unsuccessfully applied to join expeditions with Scott and Shackleton.

Mertz, who held three university degrees, was one of the best loved members of the Cape Denison team. His attempt at learning English was a source of much hilarity, being fed outrageous swear words as though they were part of every-day speech. He was also the expedition's ski instructor.

The two men formed a close bond and were the only casualties of Mawson's expedition. A memorial cross to mark their deaths was erected by their colleagues at Cape Denison, made from the mast of the wireless relay station.

A genuine fondness developed between all the men and huskies which created laughter and merriment but sometimes frustration, bewilderment, exasperation and anger with their behaviour.

During the January-February 1912 construction of what is now known as Mawson's Huts the dogs were chained together at night,

Memorial cross erected to the memory of Mertz and Ninnis
(Photograph courtysey Alisdair McGregor)

mainly to prevent them chasing the thousands of Adelie penguins and seals. They slept in the shelter of packing cases along with the men.

When completed, the two huts had a veranda around three sides, a section of which the dogs used, along with supplies and later biological specimens.

The dogs were kept well away from the building work while dynamite, warmed in trouser pockets so it could be ignited, was used to blast boulders to provide a flat surface. Long drills, worked laboriously by hand, created holes large enough to pack the sticks of dynamite with guano scraped from the rocks as packing. This blasted some of the oldest rocks in the history of the world.

Guano was also used to pack the wood footings for the huts and when mixed with the men's urine froze almost immediately, providing quick acting cement with the DNA of the inhabitants forever embedded.

Food for the dogs was plentiful, with penguins and seals in abundance during the summer. Supplies were also given a boost a month after arriving at Cape Denison with the emergence from the Southern Ocean of a giant sea elephant which lumbered ashore.

The marine monster towered over a dog Ninnis had named Johnson after a famous prize fighter and who like his namesake did not take a backward step.

The unexpected visitor, while not an apparent threat to the dogs, lumbered towards the men when they were sighted, only to be shot. Measuring 5.5 meters long and 3.65 meters around the girth it weighed many tons and provided not only blubber for the stove for use during the winter but over a ton of meat for the dogs.

This was stored in a labyrinth of snow tunnels which were natural fridges. A storage area was also created under the hut where carcasses of mutton were placed after the fleece and skin had been removed. This was tacked to the exterior of the main building and held in place with battens to keep out the constant wind. Fragments of this skin remain under the strips of wood today.

Extracting a sheep carcass was difficult in such a confined space so sometimes a husky was sent in to retrieve one for the evening meal. The trick, however, was to catch the dog when it emerged with dinner in its jaws. With food in the offing they were fast and on more than one occasion the dogs won, earning themselves a change in diet.

The huskies continued to amaze, intrigue and fascinate the men but not when there were moments of mischief such as when Shackleton ripped open a food bag and scoffed down a 1.2 kg block of precious butter.

Mertz and Ninnis harness the dogs at Cape Denison.
(Photograph Frank Hurley, courtesy of the Mitchell Library of NSW)

Ninnis and Mertz worked tirelessly to train the dogs whenever conditions allowed. The dogs were harnessed and then run up and down the nearby slopes hauling sledges. With two men singing students songs they had taught each other and yelling commands, the huskies loved racing in teams. For the dogs it was a happy time.

On the slopes 12 km above the huts Mawson created an emergency supply depot which they called Aladdin's Cave. It was an ice cave dug out with ice axes and large enough for four men to sleep in. It was so named at the suggestion of Ninnis because of the magical looking crystals hanging from the ceiling.

For some unknown reason the dogs loved it. When Ninnis and Mawson made a mid-winter attempt to trek further inland from the depot they were forced to let the dogs loose as they battled into a driving blizzard.

All but one dog turned and raced off. Pavlova, Mawson's favourite named after the Russian ballerina who Ninnis had met prior to leaving London, stayed with the men.

When they returned to Aladdin's Cave hours later the dogs were there waiting. They were rewarded with seal meat and biscuits.

Mawson and Ninnis were forced to remain inside the cave for two days while the blizzard howled and the dogs sheltered at the entrance. While cosy for the men, the two feared their absence would concern the men at base camp. As they tried to make their way down the slope with the blizzard still raging with winds well in excess of 100 kph, the dogs had difficulty with the sledge constantly tipping over.

Again they were let loose in the belief they would run with the men and sledge to the huts but they all turned and raced back to Aladdin's Cave.

Nearly a week passed before weather conditions allowed a party including Frank Hurley and Robert Bage to retrieve the dogs from the cave.

Castor (twin brother of Pollux in Greek mythology) leapt up and down with excitement but Pavlova and the other four were lethargic and sick. The men cooked the dog's hot food and nursed them for two days, but for one it signaled the end.

Grandmother, named because of his looks, died and was buried outside in the ice.

Mawson wrote in his diary on August 25:
*They made no attempt to break into the provisions, Grandmother was in the worst condition and died soon after relief the dogs were all within a day of expiring. Grandmother an hour before death, looked straight ahead with glassy eyes.*

The dogs continued their fascination with the ice cave. Weeks later Ninnis and Mertz ferried more supplies to the depot and again the dogs were let loose on the journey home. Again they simply turned and returned to the hole in the ice.

This time they were put on a trace but returned again when freed back at Mawson's Huts. Retrieved again and taken all the way back to base, two dug their way out of their ice tunnel that night and returned to the cave. Mawson commented:
*The ways of Greenland husky dogs are beyond human comprehension.*

One of them, called Franklin (named after Sir John Franklin the famous British Arctic explorer who also served as Lieutenant-Governor of Van Diemen's Land, later Tasmania), wagged his tail and snuggled down with the men. The second, Scott (named after Sir Robert Falcon Scott) crept away and was never sighted again.

Eighty six years later Grandmother's skeletal remains were retrieved from the plateau by members of the 1997-98 Mawson's Huts Foundation conservation expedition.

Mawson's Huts Foundation expedition 1997-98.
Leader Alan Hunt inspects Grandmother.
(Photograph Malcolm Ludgate)

They were on the surface of the plateau not far from Aladdin's Cave.

Author and artist Alasdair McGregor on the 1997 expedition, recorded the recovery of Grandmother in his published journey *Mawson's Huts. An Antarctic Expedition Journal.*
In January 1998 he spotted three dark specks about 150 meters from each other. He wrote:

> *Two are AAE fuel containers and the dog lies between. Aladdin's Cave must have been somewhere close at hand but with movement of the ice in the intervening years, it would be next to impossible to find any trace of it today.*

*Whether the fuel tins dropped off a sled, or were originally left at the cave, or simply thrown away, no one will ever know. Stuck to the ice by a few tufts of remaining fur on the back of its head, the dog is a piteous sight, even as a skeleton. In death it still appears to cower against the blast [wind]-resigned to exhaustion and curled in enfeebled defiance to the cold.*

*Remarkably some of its desiccated viscera remains within the body cavity, while tendons and sinews are permanently taut and teeth clenched. It strikes me as a subject to which the likes of Drysdale, Nolan or Tucker* [Australian artists] *might respond: the brutality of life and death in an alien landscape.*

The decision was made to return the dog to Mawson's Huts where it remains in a wooden box inside – the one place where dogs were never allowed.

# 8

# THOUGHTS OF A DOG HANDLER

Ninnis wrote much about the dogs in his diary before the ill-fated trek and for Tuesday March 12, 1912 he penned these personal thoughts on the characteristics of his dogs:

*I have written much of the dogs but never given each one's biography and characteristics. First and foremost of course comes* **Basilisk**, *the King dog. A big, strong solemn and middle aged beast with a fine black coat and white chest and an imposing presence. He is an ideal king; he never abuses his position and seldom bites one of the small or weaker dogs. His rule has never yet been seriously disputed which is just as well for I have never seen a dog in the same street as him for speed on onslaught. In peace he appears slow and sedate but in war he is swift and terrible, delivering a bite with lightning speed and no dogs wants another. Basilisk is Mertz's special dog.*

*   **Alexandra**, *more commonly known as* **Ginger Bitch**, *is Basilisk's greatest friend. She is a big, ginger coloured animal, by far the largest of the bitches and a tyrannical queen among them. She, like her lord and master, does not bark but gives a croak of welcome as she greets her friends. She and Basilisk are inseparable, a veritable Darby and Joan and by her jealous antics when one or other of them approaches a member of the opposite sex, causes us endless amusement.*

*   **Shackleton** *is another big black dog but with a patch of white on his chest. A boisterous affectionate beast, he is the most popular dog in the pack and quite the handsomest. From his little way of pushing against one's legs with his head and his friendly offering of his paw to shake, all his little ways are attractive. He is never ill and never cast down. In the worst weather during the voyage, as in the blizzards here,*

*his tail is always cocked jauntily over his back and drenched with salt water, panting under a tropical sun or his coat with a mass of icicles, he is always up and lively.*

*He fears no dog but Basilisk and he is the best puller in the pack, Alexandra running him close. His one weakness is his penchant for seals and penguins, but his subdued and chastened aspect when caught at his misdeeds disarms the would-be avenger and he usually escapes with the lightest of whippings.*

*__Franklin__ is not unlike Shackleton in appearance but unlike him is always gloomy in aspect and a poor and lazy puller, and altogether devoid of attractive [sic]. A good and savage fighter he commands the respect of all but Basilisk.*

*His great and only friend is __Scott__ a white dog with a black face, of no marked characteristics but a nice tame and friendly beggar, popular with most of the pack.*

*__Fusilier__ is probably the strongest dog in the pack, after the redoubtable Basilisk. His colour is very dark brown with the shorter and silky coat than the black dogs but his chest is very broad and his legs remarkable muscular, and he should if he exerts himself be one of the best pullers of all but he is a shy, timid dog, albeit friendly. His habit of greeting advances being to lie on his back with legs in the air inviting friction where he values it most. He is on terms of peace with all but his great friend is Caruso. The two are like David and Goliath.*

*__Caruso__ is another dark brown dog, lank and lean with a sharp foxy face and the most unattractive dog in the pack. He has no pleasant ways and only in Fusilier's company is he brisk and lively. He is now and always has been mangy and disreputable in appearance.*

*__George__ and __Mary__ are smallish white dogs with sharp black faces and brown eyes, being almost similar in appearance, George being a little larger of the two with a black spot on his rump. Both* *are very pretty, friendly and attractive and are great favourites with all. George is a plucky fighter, being very quick on his feet and with a lightning delivery but is rather handicapped with his weight.*

*__John Bull__ is the pariah of the pack being universally unpopular with the other dogs, he is a middle sized, strong thickset dog with a very fine black coat with a white chest like Basilisk. He never attempts to conciliate the other dogs but although friendly disposed towards*

Ginger and her pups on board *Aurora*

*human beings, snaps and snarls at all his fellow canines, a trick that earned him his name after Horatio Bottomley's famous paper.*

**Castor** *a biggish white dog with a sharp face and light brown eyes and muzzle. He is affectionate and fairly attractive and very friendly.*

*His bosom friend is* **Haldane**, *a big long wolfish dog, wolfish in colour and appearance and, at the start in disposition, being the cur of the pack and so earning his name. He is a strong animal and although excessively shy and timid at first, has plucked up wonderfully of late being much more friendly.*

**Jack Johnson** *is another of our white dogs, a pugilist as his name implies but a most unlucky one at that, being small and slow and never having never been known to win a fight until recently when Basilisk generously took him under his wing. Owing to his fighting qualities his face is scarred all over being quite devoid of fur, a fact that with mange, make him a sorry sight. His ears droop miserably, his tail hangs limp and his appearance is woebegone in the extreme affording us much amusement.*

**Grandmother** *despite his name, he being a dog, is one of our best dogs being exactly similar to Fusilier but he is much more boisterous and affectionate and a famous puller. He is very peaceful and a great favourite. He was the first dog to bite me, albeit accidentally.*

**Betli** *is a medium sized bitch, white body and a black face and a nice coat but skimpy tail. She is a demonstratively affectionate dog and a great favourite with Mertz, who named her, despite her penchant for penguin hunting. She is one of only two dogs who once having a fit, survived.*

**Pavlova** *is the handsomest, of our bitches. A pretty, lively and very affectionate animal with a fine black coat and white chest. She is one of the most popular of the pack and a plucky fighter, even venturing to square up to Alexandra.*

***Ginger*** *is a big, lively, greedy and affectionate beast (bitch) and has filled out and improved wonderfully since we had her. During the voyage she was always on the sick list, being very excitable, not to say hysterical and prone to fits, but now she is well and in fine condition. She has had one litter of nine, but being young and inexperienced, she proved a bad mother and her offspring speedily succumbed.*

***Gadget*** *is a small, mouse coloured and unattractive bitch but a good mother and she is now nursing a fine pup.*

***Jappy*** *is easily our most useless dog. Very small and wretched, she and is prone to howl at nothing at all, always shivering and miserable looking, she is disliked by all. Her size must always preclude her from being any good and food on her is merely wasted. I expect she will be got rid of.*

# 9

# THEY LIVED AND DIED TOGETHER

Basilisk with Alexandra, 1912
(Photograph Frank Hurley)

The story of Basilisk and Alexandra is one of most moving stories of devotion between animals ever witnessed in the Antarctic.

Constant companions since being selected in Greenland to be part of Mawson's AAE, the two became devoted to each other until their death. Harnessed together they plunged down a crevasse with the sledge they were hauling along with their handler, Belgrave Ninnis.

Throughout their trip to Hobart from London, during their time at the quarantine station at Taroona, on the outskirts of Hobart where several of the dogs died from distemper, Basilisk looked after Alexandra; the two were inseparable.

They slept curled up side by side at night, played together during the day and were seldom apart. At feeding time Basilisk made sure Alexandra got a little bit extra by taking some of the food from the other dogs.

This practice continued at Cape Denison with Basilisk scavenging for pieces of seal or penguin that were slaughtered to feed both the dogs and expeditioners.

His devotion to "Ginger Bitch" caused great interest and amusement to the men of the AAE who watched Basilisk running around finding little delicacies for his partner.

They had several offspring between them with one named "Blizzard" who became a favourite with all the men.

Hurley immortalised Blizzard when he took a beautiful photo of the pup sitting on the ice. This picture is now in Hurley's AAE collection held by the Mitchell Library of the NSW State Library.

Blizzard

# 10

# DOGS OF THE AUSTRALIAN NATIONAL ANTARCTIC EXPEDITION (ANARE)

## Rob Easther

*The husky dog is an important constituent of the romance and fascination of polar history. Transport in Polar Regions has always been difficult and before the advent of tracked vehicles and helicopters, dog sledging provided a safe, efficient and economical means of travelling. But the dogs were more than workers, to the people at remote polar outposts they were objects of deep affection.*
Dr Phillip Law, Director of ANARE 1949 to 1966.

On 4 November 1992, two teams of huskies bounded onto the sea-ice at Mawson station, Antarctica, heading for the Australian icebreaker *Aurora Australis*, moored 63km away at the edge of the sea ice.

Their journey marked the end of the nation's 43 year affair with the gallant, much-loved working dogs that, in their heyday, made exploration over the sea ice and Antarctic plateau possible.

Although their primary role had long diminished with the increasingly efficient modern over-snow transport, they remained at Mawson Station as a major draw card for many expeditioners who loved their companionship and the thrill of a run with the dogs. For most Mawson expeditioners, the dogs were considered a great morale booster for the long winter months.

*...given the ills that things mechanical are subject to, no Antarctic station is self-contained without the back-up of a couple of good dog teams...I will quote the 1960 example of the dogs taking*

A dogger controls the nine-dog team during a rest stop, 1986
(Photograph courtesy Matt Sherlock)

*fuel inland for tractors returning from the southern Prince Charles Mountains, and salvaging valuable equipment from a wrecked aircraft.* Syd Kirkby - dogger 1956, 1960.

A succession of devoted expeditioners kept alive the skills, equipment and training for both dogs and their handlers, a valued remnant of the earliest days of Australia's Antarctic program at Heard Island, Mawson and Davis stations and for a brief time, Casey station, inherited from nearby American station Wilkes when it closed down in 1969.

The Australian dog's bloodline goes back to an arrangement in 1949 between the French and Australian Antarctic programs whereby the French dogs remained in quarantine for 12 months in the Melbourne Zoo and pups born during their stay were offered to the Australian program and formed the basis of 12 dogs shipped to Australia's research station at Atlas Cove on Heard Island in 1950.

In February 1954, 30 huskies with the 10 wintering expeditioners they were to serve, arrived from Atlas Cove for the establishment of Australia's first Antarctic continental station, Mawson.

The combination of the Labrador huskies strength and the Greenland strains - short, thick coat and even temperament - served Australia well in the harsh winters of Mawson, and, from 1957, Davis. New blood in the form of Alaskan Malamute dogs brought to Antarctica by the Americans was introduced from Wilkes.

Over the years, the huskies travelled far inland on the Antarctic plateau on journeys with surveyors, geologists, biologists, glaciologists and others to the Prince Charles Mountains 480 km south of Mawson and to Amundsen Bay, 650 km west of the station. Regular shorter runs to the Emperor penguin colonies at Auster 60 km east of Mawson and Taylor Glacier, 80kms to the west were regularly made.

Excellent running conditions for the dog team
– all doggers riding the sledge – a rare occurrence
(Photograph courtesy Matt Sherlock)

Their special aptitude on sea-ice, when they become agitated or simply pulled to a halt refusing the handlers instructions to go on, saved many expeditioners from crossing areas of thin ice. Overnight the dog's sudden excitement has been known to warn expeditions of opening tide cracks requiring a hasty de-camping to avoid the rising sea. The handlers ignored the dogs 'advice' at their peril! However, not all handlers had such faith in their 'radar'.

*They never were mystical discoverers of crevasses and tide cracks undetectable to man and I doubt that endowing them with qualities they did not have does them any service. In fact we could have run a nice little sideline at a few dollars a time to pull dogs out of crevasses. They were most amiable and devoted, they were companionable, uncritical and amusing when often, God knows, these qualities were such a tonic and restorative.*
Syd Kirkby -dogger 1956, 1960.

Of the many tales told over the years of the dogs' relationships with their handlers, perhaps the most famous concerns the much-travelled Oscar, born on Heard Island in 1951 and veteran of Mawson, Davis and Wilkes where he died in the early 1960s.

With the Antarctic Treaty update of its environmental protection policies in 1991 (the Madrid Protocol), the requirement to remove any introduced species from the natural reserve of Antarctica, meant the remaining working dogs of the British, Argentinian and Australian programs had to be withdrawn from the continent.

Their last run to the ship was to be their longest journey, their next stop, Hobart. From there they boarded a freight plane to their new home in the USA, where they would continue to haul sleds on Arctic and North American expeditions there having been no comparable outdoor centre in Australia where they could be housed and continue living the lives they had known in Antarctica as working dogs.

On the helideck of *Aurora Australis*, 19 wooden kennels constructed by expedition carpenters, were chained down for the notoriously rough passage to Hobart. The dogs were checked by a vet and with their experienced handlers and other expeditions on board the ship they set sail on the 10 day voyage.

On their safe arrival in Hobart 'the last huskies' as they became known, were transferred, with two of their handlers who had spent the last winter with them, to a flight to the US. Following a road trip from Los Angeles, they arrived in their new homes near Ely in Minnesota with two outdoor adventure centres that specialised in dog sledding expeditions. Within a year several of the Aussie dogs had run to the North Pole and were reported as very strong pullers. Their new handlers also reported the amusing news that the male dogs, having never seen trees before, instinctively sidled up to them to pee!

Departing Twin Peaks in the David Range on the return trip to Mawson.
The sledge is fully loaded with gear after several days in the mountains.
(Photograph courtesy Matt Sherlock)

At their new homes, the dogs had never had it better. Instead of lying curled up in the snow through winter as they had always done in Antarctica, they now enjoyed the luxury of individual kennels.

In fact, the very last ANARE dogs to come out of Antarctica had stayed on at Mawson for the next winter and arrived in Hobart on *Aurora Australis* on 20th December 1993. These five dogs were considered too old to continue as working dogs, and so were retired to live with former Antarctic expeditioners who had recent experience as dog handlers.

*Who was it who was drawn to those lads in the dark of the year when home seemed a universe and an age away to sit in silence with them. In those times when they see you coming, but they read your walk and your face, and they know you have come to sit and be silent, so they don't leap, they don't shout; and you are followed to their midst by those many earnest eyes, which understand. It was the dogman.*
Tom Maggs, dogger 1977, 1980.

# FURTHER INFORMATION

*Huskies in Harness – A Love story in Antarctica*
Edited by Shelagh Robinson. 1995, Kangaroo Press. A very fine collection of stories by expeditioners through the ages of adventures with the ANARE* dogs

The book and the film *The Last Husky* record the transfer of the ANARE dogs from Antarctica to the USA.

Oscar's story is told in Nils Lied's *Oscar: the true story of a husky* (1987).

*Australian National Antarctic Research Expeditions (ANARE), earlier name for the current Australian Antarctic program

# FURTHER READING

*Aurora, Douglas Mawson and the Australasian Antarctic Expedition 1911-1914.* Riffenburgh, Beau. 2011. Erskine Press, Norwich, Norfolk.

*Mawson's Forgotten Men - The 1911-1913 Antarctic Diary of Charles Turnbull Harrison.* Rossiter, Heather (ed). 1911 Pier 9, Sydney, Australia.

*The Home of the Blizzard.* Mawson, Douglas. 1915. JB Lippincott, Philadelphia.

*Mertz & 1... The Antarctic Diary of Belgrave Edward Sutton Ninnis.* Mornament, Allan & Riffenburgh, Beau (eds) 2014 Erskine Press, Norwich, Norfolk.

The *Antarctic Diaries of Andrew Dougald Watson & Alexander Lorimer Kennedy.* Riffenburgh, Beau (ed) 2018 Erskine Press, Norwich Norfolk.

# ACKNOWLEDGEMENTS

Most of the information in this book has been drawn from the diaries of men who accompanied Dr Douglas Mawson on his Australasian Antarctic Expedition of 1911-14 and the records, diaries and books on other expeditions of the heroic era of exploration between 1898 and 1916.

Dr Google was also a great help.

Surprisingly not a great deal has been written about the fate of all the huskies which were used during these early Antarctic expeditions with the exception of those of Mawson.

Details of their lives after returning are scarce but perhaps this book will help in unveiling stories on some of these wonderful brave animals that played such an important part in the early exploration of the Antarctic.

I'm very grateful to the Australian Antarctic Division's wonderful series of websites which included Wikipedia, the Scott Polar Research Institute at Cambridge UK, Canterbury Museum, Christchurch, NZ, the Mitchell Library of the NSW State Library, the South Australian Museum which houses the Mawson Collection and the New Zealand Antarctic Heritage Trust. Very knowledgeable people such as Andrew Jackson, Rod and Jeannie Ledingham, Rob Easther, Bill Burch, Stephen Grimsley, Sid Kirkby also contributed in various ways. Thanks also for Carole Borsboom for her help in proof reading and sourcing of images, some of which unfortunately I have been unable to attribute.

Books such Douglas Mawson's "Home of the Blizzard" and his wife Paquita's book, "Mawson of the Antarctic" were wonderful references. Particular thanks to Beau Riffenburgh whose essay on the dogs of Mawson's 1911-14 expedition was published by the Scott Polar Research Institute in 2012. This provided some valuable information as did the diaries of some of Mawson's men including

Blake, Charles Harrisson, Cecil (C.T) Madigan, Archibald McLean, Morton Moyes, Xavier Mertz, Belgrave Ninnis, Charles Sandell, Watson and Frank Wild.

Heather Rossiter's "Mawson's Forgotten Men," the published diaries of Charles Turnbull Harrisson, a member of the AAE's Western Party which returned with two dogs, one of which Harrisson adopted was also a useful source.

My thanks also to the Mitchell Library of the NSW State Library which houses the largest collection of images taken by Frank Hurley, Mawson's official photographer in 1911-14. And also to Mark Pharoah, curator of the Mawson collection at the Museum of South Australia for his endless support and assistance with images and documents.

Several of Hurley's remarkable photographs taken at Cape Denison have been used to help illustrate this booklet.

Most importantly my warmest thanks to my wife Lindsie for her tireless help with researching material for this book. As always she has made my efforts in compiling material for this book a lot easier.

# ABOUT THE AUTHOR

David Jensen AM is Chairman of the Mawson's Huts Foundation, a not for profit charity he founded in 1997 to conserve the historic huts at Cape Denison, East Antarctica. These historic buildings were used as the main base for two years by the Australasian Antarctic Expedition led by Dr Douglas Mawson.

David has written several books on the huts and the men who accompanied Mawson and was made a Member of the Order of Australia in 2001 for his efforts in conserving the historic site through the Foundation.

Born in New Zealand, David learned about the feats of British explorers Sir Robert Falcon Scott and Sir Ernest Shackleton early in life but knew nothing about Mawson until arriving in Australia in late 1969 as a journalist.

David regards Mawson as one of the great explorers and is continually surprised that so few of his fellow Australians know of his expeditions and his legacy which includes Australia's territorial claim to 42 per cent of the Antarctic.

The Foundation built and operates the award winning Mawson's Huts Replica Museum and also initiated the Australian Antarctic Festival which is held in the Tasmanian capital every two years. This four day event helps to promote Hobart as a gateway to the white continent, knowledge of Mawson and Australia's work and role in the Antarctic. It is also aimed at promoting the city's rich Antarctic history.

# THE MAWSON'S HUTS FOUNDATION

Established in 1997 it was formed expressly to conserve the historic huts at Cape Denison, East Antarctica, which were used by the 1911-14 Australian Antarctic Expedition as its main base for two years.

These fragile buildings are quite rare: just one of six surviving sites of the 'heroic era' of Antarctic exploration. Situated 2730 kms directly south of Hobart, they are regarded as the birthplace of Australia's Antarctic history. They also sit at what is officially the windiest place on earth at sea level where gusts of up to 350 kph have been recorded. The average daily wind strength is just over 70 kph.

Between 1997 and 2015 the Foundation has organised and financed 13 major expeditions to the site. To assist with funding the expeditions a full scale replica of the huts have been built on the Hobart waterfront, just 200 metres from where Mawson departed in December 1911.

For further information go to WWW.Mawsons-huts.org.au or email info@mawsons-huts.org.au

Scott's *Discovery* Expedition 1901-04
Drawings by Edward Wilson